MW00950540

CONTINUED FROM NOTEBOOK NO. _____ CONTINUE _____

THIS NOTEBOOK BELONGS TO: _____

SIGNATURE: _____ DATE: _____

PERSONAL INFORMATION

PHONE NUMBER: _____ EMAIL ADDRESS: _____

CITY: _____ STATE: _____ ZIP: _____

NOTES

NOTEBOOK COMPLETED ON: _____ NUMBER OF PAGE FILLED IN: _____

TABLE OF CONTENTS

PAGE	SUBJECT TITLE	DATE
1		
2		
3		
4		
5		
6		
7		
8		
9		
10		
11		
12		
13		
14		
15		
16		
17		
18		
19		
20		
21		
22		
23		
24		
25		
26		
27		
28		
29		

BOOK NO. _____

TABLE OF CONTENTS

PAGE	SUBJECT TITLE	DATE
30		
31		
32		
33		
34		
35		
36		
37		
38		
39		
40		
41		
42		
43		
44		
45		
46		
47		
48		
49		
50		
51		
52		
53		
54		
55		
56		
57		
58		

BOOK NO. _____

TABLE OF CONTENTS

PAGE	SUBJECT TITLE	DATE
59		
60		
61		
62		
63		
64		
65		
66		
67		
68		
69		
70		
71		
72		
73		
74		
75		
76		
77		
78		
79		
80		
81		
82		
83		
84		
85		
86		
87		

BOOK NO. _____

TABLE OF CONTENTS

PAGE	SUBJECT TITLE	DATE
88		
89		
90		
91		
92		
93		
94		
95		
96		
97		
98		
99		
100		
101		
102		
103		
104		
105		
106		
107		
108		
109		
110		
111		
112		
113		
114		
115		

BOOK NO. _____

Continued from page

1

5

10

15

20

25

30

35

Continued to page

SINGNATURE

DATE

DISCLOSED TO AND UNDERSTOOD BY   DATE

PROPRIETARY INFORMATION

Continued from page

2

5

10

15

20

25

30

35

Continued to page

SIGNATURE   DATE

DISCLOSED TO AND UNDERSTOOD BY   DATE

PROPRIETARY INFORMATION

Continued from page

3

5

10

15

20

25

30

35

Continued to page

SINGNATURE   DATE

DISCLOSED TO AND UNDERSTOOD BY   DATE

PROPRIETARY INFORMATION

TITLE

PROJECT

Continued from page

4

5

10

15

20

25

30

35

Continued to page

SINGNATURE

DATE

DISCLOSED TO AND UNDERSTOOD BY

DATE

PROPRIETARY INFORMATION

Continued from page

5

10

15

20

25

30

35

Continued to page

SINGNATURE

DATE

DISCLOSED TO AND UNDERSTOOD BY

DATE

PROPRIETARY INFORMATION

6

5

10

15

20

25

30

35

Continued to page

SIGNATURE   DATE

DISCLOSED TO AND UNDERSTOOD BY   DATE

PROPRIETARY INFORMATION

Continued from page

5

10

15

20

25

30

35

Continued to page

SINGNATURE   DATE

DISCLOSED TO AND UNDERSTOOD BY   DATE

PROPRIETARY INFORMATION

TITLE   PROJECT

Continued from page

8

5

10

15

20

25

30

35

Continued to page

SINGNATURE   DATE

DISCLOSED TO AND UNDERSTOOD BY   DATE

PROPRIETARY INFORMATION

Continued from page

9

5

10

15

20

25

30

35

Continued to page

SINGNATURE   DATE

DISCLOSED TO AND UNDERSTOOD BY   DATE

PROPRIETARY INFORMATION

Continued from page

10

5

10

15

20

25

30

35

Continued to page

SIGNATURE   DATE

DISCLOSED TO AND UNDERSTOOD BY   DATE

PROPRIETARY INFORMATION

Continued from page

11

5

10

15

20

25

30

35

Continued to page

SINGNATURE   DATE

DISCLOSED TO AND UNDERSTOOD BY   DATE

PROPRIETARY INFORMATION

Continued from page

5

10

15

20

25

30

35

Continued to page

SIGNATURE

DATE

DISCLOSED TO AND UNDERSTOOD BY

DATE

PROPRIETARY INFORMATION

Continued from page

Continued to page

SINGNATURE

DATE

DISCLOSED TO AND UNDERSTOOD BY

DATE

PROPRIETARY INFORMATION

Continued from page

5

10

15

20

25

30

35

Continued to page

SIGNATURE

DATE

DISCLOSED TO AND UNDERSTOOD BY

DATE

PROPRIETARY INFORMATION

Continued from page

Continued to page

SINGNATURE

DATE

DISCLOSED TO AND UNDERSTOOD BY

DATE

PROPRIETARY INFORMATION

Continlued from page

5

10

15

20

25

30

35

Continued to page

Continued from page

5

10

15

20

25

30

35

Continued to page

SINGNATURE

DATE

DISCLOUSED TO AND UNDERSTOOD BY

DATE

PROPRIETARY INFORMATION

Continued from page

18

5

10

15

20

25

30

35

Continued to page

SIGNATURE

DATE

DISCLOSED TO AND UNDERSTOOD BY

DATE

PROPRIETARY INFORMATION

TITLE   PROJECT

Continued from page

19

5

10

15

20

25

30

35

Continued to page

SINGNATURE   DATE

DISCLOSED TO AND UNDERSTOOD BY   DATE

PROPRIETARY INFORMATION

Continued from page

20

5

10

15

20

25

30

35

Continued to page

INGNATURE

DATE

ISCLOSED TO AND UNDERSTOOD BY

DATE

PROPRIETARY INFORMATION

Continued from page

Continued to page

SINGNATURE   DATE

DISCLOSED TO AND UNDERSTOOD BY   DATE

PROPRIETARY INFORMATION

5

10

15

20

25

30

35

Continued to page

SIGNATURE

DATE

DISCLOSED TO AND UNDERSTOOD BY

DATE

PROPRIETARY INFORMATION

Continued from page

23

5

10

15

20

25

30

35

Continued to page

SINGNATURE   DATE

DISCLOSED TO AND UNDERSTOOD BY   DATE

PROPRIETARY INFORMATION

Continued from page

24

5

10

15

20

25

30

35

Continued to page

INGNATURE

DATE

ISCLOSED TO AND UNDERSTOOD BY

DATE

PROPRIETARY INFORMATION

Continued from page

25

Continued to page

SINGNATURE   DATE

DISCLOSED TO AND UNDERSTOOD BY   DATE

PROPRIETARY INFORMATION

continued from page

26

5

10

15

20

25

30

35

Continued to page

SIGNATURE   DATE

DISCLOSED TO AND UNDERSTOOD BY   DATE

PROPRIETARY INFORMATION

TITLE

PROJECT

Continued from page

27

5

10

15

20

25

30

35

Continued to page

SINGNATURE

DATE

DISCLOSED TO AND UNDERSTOOD BY

DATE

PROPRIETARY INFORMATION

Continued from page

5

10

15

20

25

30

35

Continued to page

INGNATURE

DATE

ISCLOSED TO AND UNDERSTOOD BY

DATE

PROPRIETARY INFORMATION

Continued from page

29

5

10

15

20

25

30

35

Continued to page

SINGNATURE

DATE

DISCLOSED TO AND UNDERSTOOD BY

DATE

PROPRIETARY INFORMATION

continued from page

5

10

15

20

25

30

35

Continued to page

SIGNATURE

DATE

DISCLOSED TO AND UNDERSTOOD BY

DATE

PROPRIETARY INFORMATION

Continued from page

5

10

15

20

25

30

35

Continued to page

SINGNATURE

DATE

DISCLOSED TO AND UNDERSTOOD BY

DATE

PROPRIETARY INFORMATION

Continued from page

32

5

10

15

20

25

30

35

Continued to page

SIGNATURE

DATE

DISCLOSED TO AND UNDERSTOOD BY

DATE

PROPRIETARY INFORMATION

Continued from page

33

5

10

15

20

25

30

35

Continued to page

SINGNATURE   DATE

DISCLOSED TO AND UNDERSTOOD BY   DATE

PROPRIETARY INFORMATION

continued from page

5

10

15

20

25

30

35

Continued to page

SIGNATURE

DATE

DISCLOSED TO AND UNDERSTOOD BY

DATE

PROPRIETARY INFORMATION

Continued from page

35

5

10

15

20

25

30

35

Continued to page

SINGNATURE   DATE

DISCLOSED TO AND UNDERSTOOD BY   DATE

PROPRIETARY INFORMATION

TITLE

PROJECT

Continued from page

36

5

10

15

20

25

30

35

Continued to page

SIGNATURE

DATE

DISCLOSED TO AND UNDERSTOOD BY

DATE

PROPRIETARY INFORMATION

Continued from page

37

5

10

15

20

25

30

35

Continued to page

SINGNATURE

DATE

DISCLOSED TO AND UNDERSTOOD BY

DATE

PROPRIETARY INFORMATION

ontinued from page

38

5

10

15

20

25

30

35

Continued to page

GNATURE   DATE

SCLOSED TO AND UNDERSTOOD BY   DATE

PROPRIETARY INFORMATION

Continued from page

39

5

10

15

20

25

30

35

Continued to page

SINGNATURE

DATE

DISCLOSED TO AND UNDERSTOOD BY

DATE

PROPRIETARY INFORMATION

Continued from page

40

5

10

15

20

25

30

35

Continued to page

INGNATURE

DATE

ISCLOSED TO AND UNDERSTOOD BY

DATE

PROPRIETARY INFORMATION

Continued from page

41

5

10

15

20

25

30

35

Continued to page

SINGNATURE   DATE

DISCLOSED TO AND UNDERSTOOD BY   DATE

PROPRIETARY INFORMATION

ontinued from page

5

10

15

20

25

30

35

Continued to page

GNATURE

DATE

SCLOSED TO AND UNDERSTOOD BY

DATE

PROPRIETARY INFORMATION

Continued from page

5

10

15

20

25

30

35

Continued to page

SINGNATURE

DATE

DISCLOSED TO AND UNDERSTOOD BY

DATE

PROPRIETARY INFORMATION

Continued from page

5

10

15

20

25

30

35

Continued to page

INGNATURE

DATE

ISCLOSED TO AND UNDERSTOOD BY

DATE

PROPRIETARY INFORMATION

Continued from page

5

10

15

20

25

30

35

Continued to page

SINGNATURE

DATE

DISCLOSED TO AND UNDERSTOOD BY

DATE

PROPRIETARY INFORMATION

Continued from page

5

10

15

20

25

30

35

Continued to page

SIGNATURE

DATE

DISCLOSED TO AND UNDERSTOOD BY

DATE

PROPRIETARY INFORMATION

Continued from page

5

10

15

20

25

30

35

Continued to page

SINGNATURE

DATE

DISCLOSED TO AND UNDERSTOOD BY

DATE

PROPRIETARY INFORMATION

TITLE   PROJECT

Continued from page

48

5

10

15

20

25

30

35

Continued to page

SIGNATURE   DATE

DISCLOSED TO AND UNDERSTOOD BY   DATE

PROPRIETARY INFORMATION

TITLE   PROJECT

Continued from page

49

5

10

15

20

25

30

35

Continued to page

SINGNATURE   DATE

DISCLOSED TO AND UNDERSTOOD BY   DATE

PROPRIETARY INFORMATION

Continued from page

5

10

15

20

25

30

35

Continued to page

SIGNATURE

DATE

DISCLOSED TO AND UNDERSTOOD BY

DATE

PROPRIETARY INFORMATION

Continued from page

5:

5

10

15

20

25

30

35

Continued to page

SINGNATURE   DATE

DISCLOSED TO AND UNDERSTOOD BY   DATE

PROPRIETARY INFORMATION

Continlued from page

5

10

15

20

25

30

35

Continued to page

SIGNATURE

DATE

DISCLOSED TO AND UNDERSTOOD BY

DATE

PROPRIETARY INFORMATION

Continued from page

Continued to page

SINGNATURE

DATE

DISCLOSED TO AND UNDERSTOOD BY

DATE

PROPRIETARY INFORMATION

Continued from page

54

5

10

15

20

25

30

35

Continued to page

SIGNATURE   DATE

DISCLOSED TO AND UNDERSTOOD BY   DATE

PROPRIETARY INFORMATION

Continued from page

5

Continued to page

SINGNATURE   DATE

DISCLOSED TO AND UNDERSTOOD BY   DATE

PROPRIETARY INFORMATION

Continued from page

56

5

10

15

20

25

30

35

Continued to page

SIGNATURE

DATE

ISCLOSED TO AND UNDERSTOOD BY

DATE

PROPRIETARY INFORMATION

Continued from page

57

5

10

15

20

25

30

35

Continued to page

SINGNATURE

DATE

DISCLOSED TO AND UNDERSTOOD BY

DATE

PROPRIETARY INFORMATION

Continued from page

5

10

15

20

25

30

35

Continued to page

SIGNATURE

DATE

DISCLOSED TO AND UNDERSTOOD BY

DATE

PROPRIETARY INFORMATION

Continued from page

59

5

10

15

20

25

30

35

Continued to page

SINGNATURE   DATE

DISCLOSED TO AND UNDERSTOOD BY   DATE

PROPRIETARY INFORMATION

Continued from page

5

10

15

20

25

30

35

Continued to page

SIGNATURE

DATE

DISCLOSED TO AND UNDERSTOOD BY

DATE

PROPRIETARY INFORMATION

Continued from page

61

5

10

15

20

25

30

35

Continued to page

SINGNATURE   DATE

DISCLOSED TO AND UNDERSTOOD BY   DATE

PROPRIETARY INFORMATION

ontinued from page

5

10

15

20

25

30

35

Continued to page

GNATURE

DATE

CLOSED TO AND UNDERSTOOD BY

DATE

PROPRIETARY INFORMATION

Continued from page

63

5

10

15

20

25

30

35

Continued to page

SINGNATURE

DATE

DISCLOSED TO AND UNDERSTOOD BY

DATE

PROPRIETARY INFORMATION

Continued from page

5

10

15

20

25

30

35

Continued to page

SIGNATURE

DATE

DISCLOSED TO AND UNDERSTOOD BY

DATE

PROPRIETARY INFORMATION

Continued from page

65

5

10

15

20

25

30

35

Continued to page

SINGNATURE

DATE

DISCLOSED TO AND UNDERSTOOD BY

DATE

PROPRIETARY INFORMATION

ntinued from page

5

10

15

20

25

30

35

Continued to page

GNATURE

DATE

CLOSED TO AND UNDERSTOOD BY

DATE

PROPRIETARY INFORMATION

Continued from page

5

10

15

20

25

30

35

Continued to page

SINGNATURE

DATE

DISCLOSED TO AND UNDERSTOOD BY

DATE

PROPRIETARY INFORMATION

5

10

15

20

25

30

35

Continued to page

SIGNATURE

DATE

DISCLOSED TO AND UNDERSTOOD BY

DATE

PROPRIETARY INFORMATION

TITLE

PROJECT

Continued from page

69

5

10

15

20

25

30

35

Continued to page

SINGNATURE

DATE

DISCLOSED TO AND UNDERSTOOD BY

DATE

PROPRIETARY INFORMATION

ntinued from page

70

5

10

15

20

25

30

35

Continued to page

GNATURE

DATE

CLOSED TO AND UNDERSTOOD BY

DATE

PROPRIETARY INFORMATION

Continued from page

Continued to page

SINGNATURE | DATE

DISCLOSED TO AND UNDERSTOOD BY | DATE

PROPRIETARY INFORMATION

Continued from page

Continued to page

SIGNATURE

DATE

DISCLOSED TO AND UNDERSTOOD BY

DATE

PROPRIETARY INFORMATION

Continued from page

5

10

15

20

25

30

35

Continued to page

SINGNATURE   DATE

DISCLOSED TO AND UNDERSTOOD BY   DATE

PROPRIETARY INFORMATION

ntinued from page

5

10

15

20

25

30

35

Continued to page

GNATURE

DATE

CLOSED TO AND UNDERSTOOD BY

DATE

PROPRIETARY INFORMATION

Continued from page

7.

Continued to page

SINGNATURE   DATE

DISCLOSED TO AND UNDERSTOOD BY   DATE

PROPRIETARY INFORMATION

Continued from page

5

10

15

20

25

30

35

Continued to page

SIGNATURE

DATE

DISCLOSED TO AND UNDERSTOOD BY

DATE

PROPRIETARY INFORMATION

Continued from page

5

10

15

20

25

30

35

Continued to page

SINGNATURE

DATE

DISCLOSED TO AND UNDERSTOOD BY

DATE

PROPRIETARY INFORMATION

ntinued from page

78

5

10

15

20

25

30

35

Continued to page

GNATURE

DATE

CLOSED TO AND UNDERSTOOD BY

DATE

PROPRIETARY INFORMATION

Continued from page

7

5

10

15

20

25

30

35

Continued to page

SINGNATURE   DATE

DISCLOSED TO AND UNDERSTOOD BY   DATE

PROPRIETARY INFORMATION

Continued from page

5

10

15

20

25

30

35

Continued to page

SIGNATURE   DATE

DISCLOSED TO AND UNDERSTOOD BY   DATE

PROPRIETARY INFORMATION

Continued from page

81

5

10

15

20

25

30

35

Continued to page

SINGNATURE　　　　　　　　　　　　　　　　DATE

DISCLOSED TO AND UNDERSTOOD BY　　　　　DATE

PROPRIETARY INFORMATION

ntinued from page

5

10

15

20

25

30

35

Continued to page

GNATURE

DATE

CLOSED TO AND UNDERSTOOD BY

DATE

PROPRIETARY INFORMATION

Continued from page

8

5

10

15

20

25

30

35

Continued to page

SINGNATURE   DATE

DISCLOSED TO AND UNDERSTOOD BY   DATE

PROPRIETARY INFORMATION

Continued from page

5

10

15

20

25

30

35

Continued to page

SIGNATURE

DATE

DISCLOSED TO AND UNDERSTOOD BY

DATE

PROPRIETARY INFORMATION

Continued from page

85

Continued to page

SINGNATURE   DATE

DISCLOSED TO AND UNDERSTOOD BY   DATE

PROPRIETARY INFORMATION

ntinued from page

86

5

10

15

20

25

30

35

Continued to page

GNATURE

DATE

CLOSED TO AND UNDERSTOOD BY

DATE

PROPRIETARY INFORMATION

Continued from page

8

5

10

15

20

25

30

35

Continued to page

SINGNATURE   DATE

DISCLOSED TO AND UNDERSTOOD BY   DATE

PROPRIETARY INFORMATION

Continued from page

88

5

10

15

20

25

30

35

Continued to page

SIGNATURE   DATE

DISCLOSED TO AND UNDERSTOOD BY   DATE

PROPRIETARY INFORMATION

Continued from page

89

5

10

15

20

25

30

35

Continued to page

SINGNATURE

DATE

DISCLOSED TO AND UNDERSTOOD BY

DATE

PROPRIETARY INFORMATION

ntinued from page

90

5

10

15

20

25

30

35

Continued to page

GNATURE

DATE

CLOSED TO AND UNDERSTOOD BY

DATE

PROPRIETARY INFORMATION

Continued from page

Continued to page

SINGNATURE

DATE

DISCLOSED TO AND UNDERSTOOD BY

DATE

Continued from page

5

10

15

20

25

30

35

Continued to page

SIGNATURE

DATE

DISCLOSED TO AND UNDERSTOOD BY

DATE

PROPRIETARY INFORMATION

Continued from page

93

5

10

15

20

25

30

35

Continued to page

SINGNATURE   DATE

DISCLOSED TO AND UNDERSTOOD BY   DATE

PROPRIETARY INFORMATION

ntinued from page

94

5

10

15

20

25

30

35

Continued to page

GNATURE   DATE

CLOSED TO AND UNDERSTOOD BY   DATE

PROPRIETARY INFORMATION

Continued from page

Continued to page

SINGNATURE.

DATE

DISCLOSED TO AND UNDERSTOOD BY

DATE

PROPRIETARY INFORMATION

Continued from page

96

5

10

15

20

25

30

35

Continued to page

SIGNATURE   DATE

DISCLOSED TO AND UNDERSTOOD BY   DATE

PROPRIETARY INFORMATION

Continued from page

97

5

10

15

20

25

30

35

Continued to page

SINGNATURE

DATE

DISCLOSED TO AND UNDERSTOOD BY

DATE

PROPRIETARY INFORMATION

ntinued from page

5

10

15

20

25

30

35

Continued to page

GNATURE

DATE

CLOSED TO AND UNDERSTOOD BY

DATE

Continued from page

9

5

10

15

20

25

30

35

Continued to page

SINGNATURE   DATE

DISCLOSED TO AND UNDERSTOOD BY   DATE

PROPRIETARY INFORMATION

Continued from page

5

10

15

20

25

30

35

Continued to page

SIGNATURE

DATE

DISCLOSED TO AND UNDERSTOOD BY

DATE

PROPRIETARY INFORMATION

Continued from page

5

10

15

20

25

30

35

Continued to page

SINGNATURE

DATE

DISCLOSED TO AND UNDERSTOOD BY

DATE

PROPRIETARY INFORMATION

ntinued from page

5

10

15

20

25

30

35

Continued to page

GNATURE   DATE

CLOSED TO AND UNDERSTOOD BY   DATE

PROPRIETARY INFORMATION

TITLE   PROJECT

Continued from page

10

5

10

15

20

25

30

35

Continued to page

SINGNATURE   DATE

DISCLOSED TO AND UNDERSTOOD BY   DATE

PROPRIETARY INFORMATION

Continued from page

5

10

15

20

25

30

35

Continued to page

SIGNATURE

DATE

DISCLOSED TO AND UNDERSTOOD BY

DATE

PROPRIETARY INFORMATION

Continued from page

5

10

15

20

25

30

35

Continued to page

SINGNATURE

DATE

DISCLOSED TO AND UNDERSTOOD BY

DATE

PROPRIETARY INFORMATION

ntinued from page

5

10

15

20

25

30

35

Continued to page

GNATURE

DATE

CLOSED TO AND UNDERSTOOD BY

DATE

PROPRIETARY INFORMATION

TITLE   PROJECT

Continued from page

10

5

10

15

20

25

30

35

Continued to page

SINGNATURE   DATE

DISCLOSED TO AND UNDERSTOOD BY   DATE

PROPRIETARY INFORMATION

Continued from page

5

10

15

20

25

30

35

Continued to page

SIGNATURE

DATE

DISCLOSED TO AND UNDERSTOOD BY

DATE

PROPRIETARY INFORMATION

TITLE

PROJECT

Continued from page

10⁣

5

10

15

20

25

30

35

Continued to page

SINGNATURE

DATE

DISCLOSED TO AND UNDERSTOOD BY

DATE

PROPRIETARY INFORMATION

Continued from page

11

5

10

15

20

25

30

35

Continued to page

SINGNATURE   DATE

DISCLOSED TO AND UNDERSTOOD BY   DATE

PROPRIETARY INFORMATION

Continued from page

112

5

10

15

20

25

30

35

Continued to page

SIGNATURE   DATE

DISCLOSED TO AND UNDERSTOOD BY   DATE

PROPRIETARY INFORMATION

TITLE

PROJECT

Continued from page

11.

5

10

15

20

25

30

35

Continued to page

SINGNATURE

DATE

DISCLOSED TO AND UNDERSTOOD BY

DATE

PROPRIETARY INFORMATION

ntinued from page

114

5

10

15

20

25

30

35

Continued to page

GNATURE

DATE

CLOSED TO AND UNDERSTOOD BY

DATE

PROPRIETARY INFORMATION

Continued from page

11

5

10

15

20

25

30

35

Continued to page

SINGNATURE   DATE

DISCLOSED TO AND UNDERSTOOD BY   DATE

PROPRIETARY INFORMATION

Made in the USA
Las Vegas, NV
10 January 2023

65376151R00070